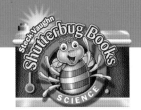

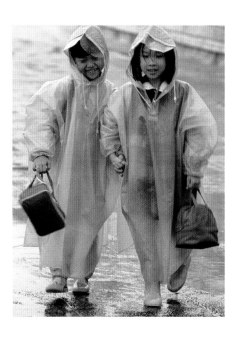

We Need Rain

by Lisa Trumbauer

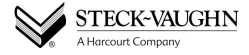

STECK-VAUGHN
A Harcourt Company

www.steck-vaughn.com

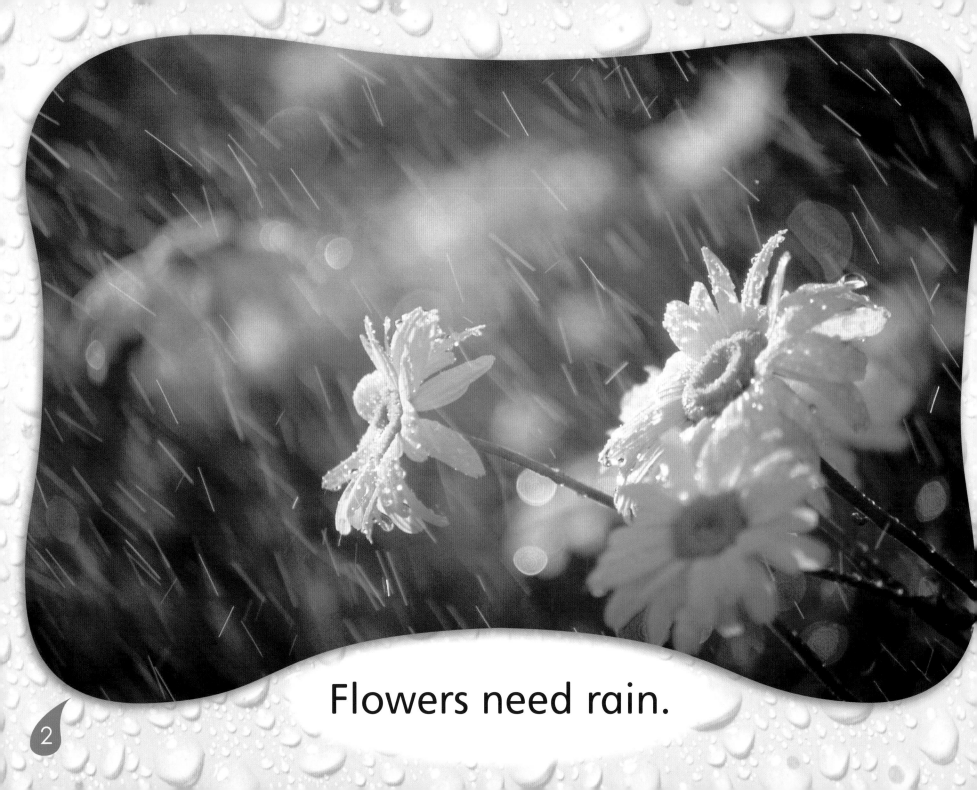

Flowers need rain.

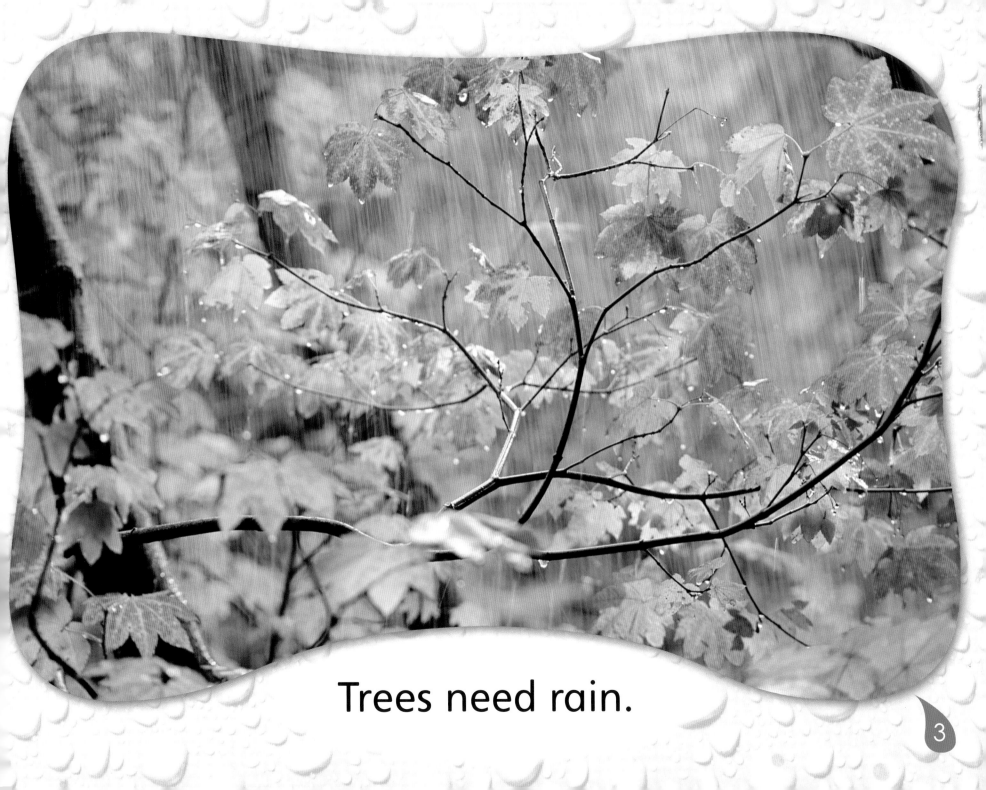

Trees need rain.

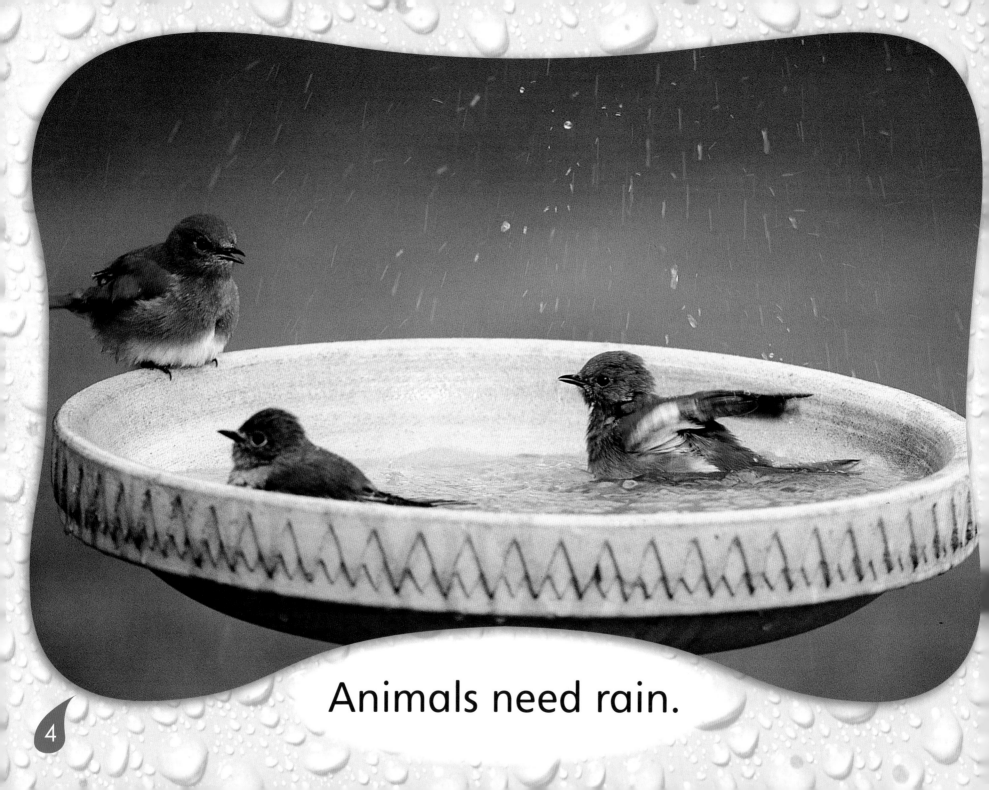

Animals need rain.

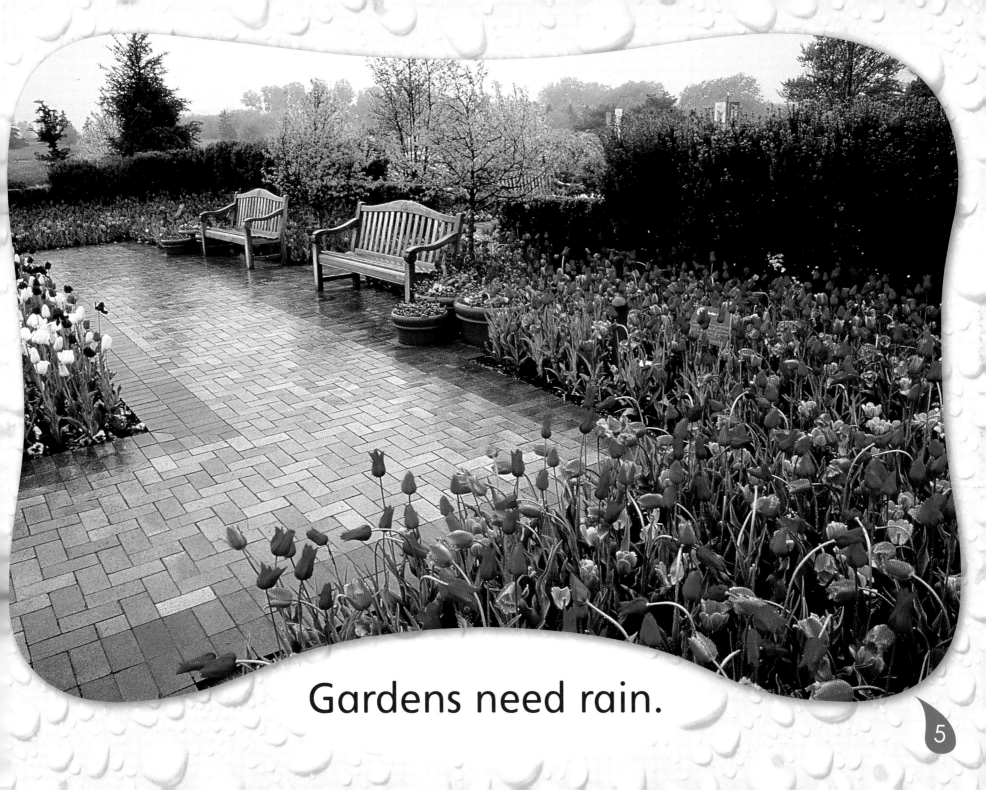

Gardens need rain.

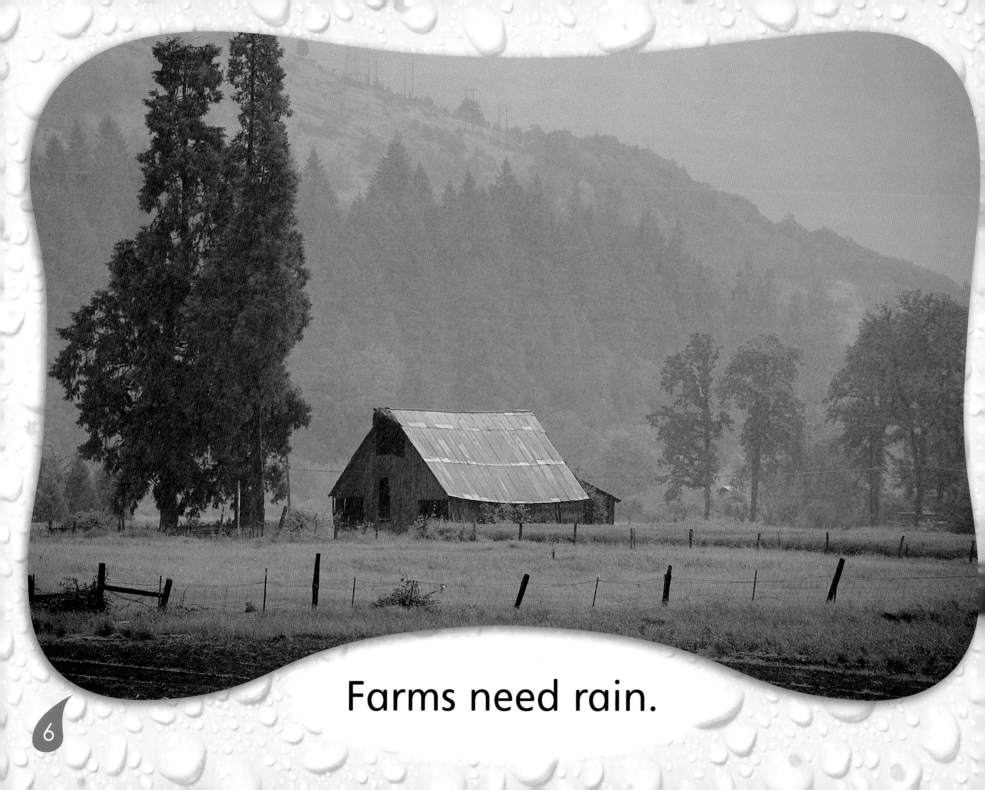

Farms need rain.

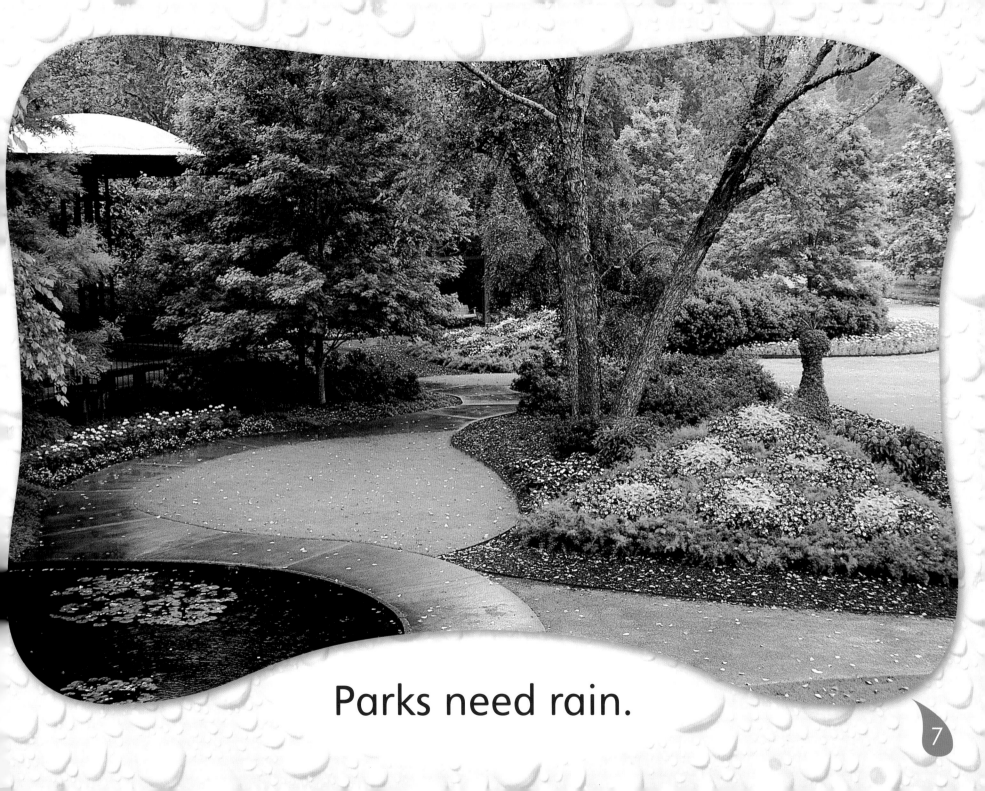

Parks need rain.

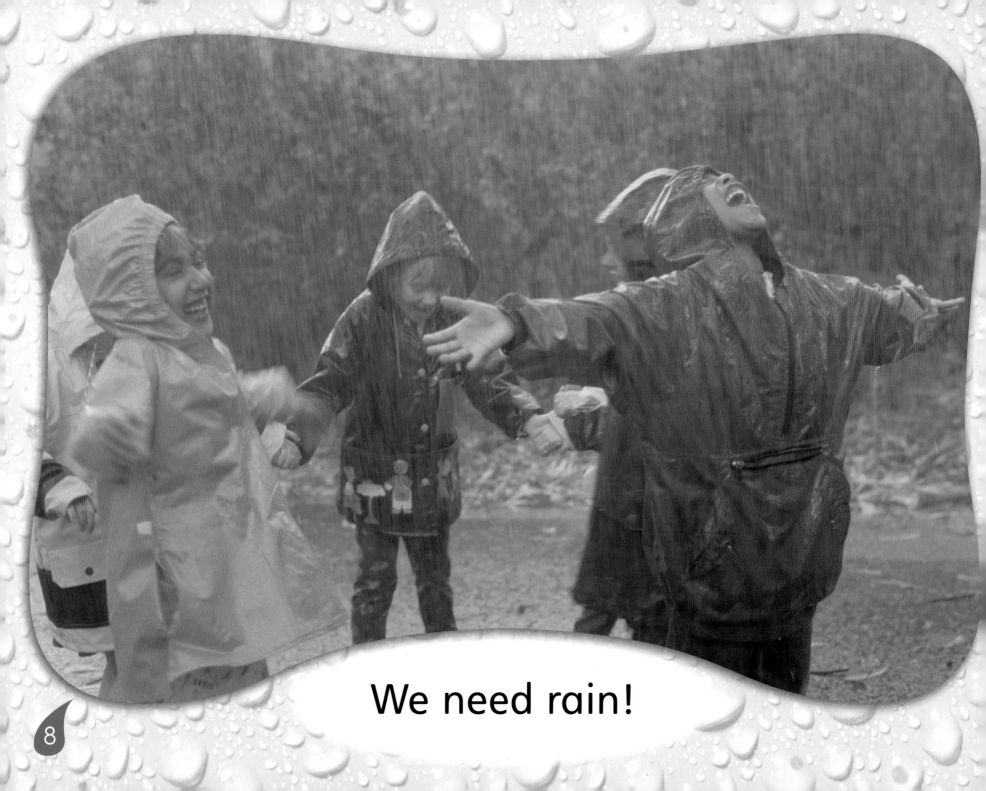

We need rain!